Grade 4

California
Unit 4
Earthquakes

Discovery Education | SCIENCE TECHBOOK

Copyright © 2019 by Discovery Education, Inc. All rights reserved. No part of this work may be reproduced, distributed, or transmitted in any form or by any means, or stored in a retrieval or database system, without the prior written permission of Discovery Education, Inc.

NGSS is a registered trademark of Achieve. Neither Achieve nor the lead states and partners that developed the Next Generation Science Standards were involved in the production of this product, and do not endorse it.

To obtain permission(s) or for inquiries, submit a request to:

Discovery Education, Inc.
4350 Congress Street, Suite 700
Charlotte, NC 28209
800-323-9084
Education_Info@DiscoveryEd.com

ISBN 13: 978-1-68220-548-8

Printed in the United States of America.

3 4 5 6 7 8 9 10 CWM 27 26 25 24 23 22   B

© Discovery Education | www.discoveryeducation.com

Acknowledgments

Acknowledgment is given to photographers, artists, and agents for permission to feature their copyrighted material.

Cover and inside cover art: austinding / Shutterstock.com

Table of Contents

Unit 4: Earthquakes

Letter to the Parent/Guardian vi

Unit Overview 1

   Anchor Phenomenon: A Bridge in an Earthquake 2

Unit Project Preview: Designing a Safe Bridge 4

Concept 4.1 Earthquake Causes and Impacts

Concept Overview 6

   Wonder 8

      Investigative Phenomenon: Experiencing an Earthquake 10

   Learn 16

   Share 38

Concept 4.2 Earthquake Waves

Concept Overview 48

   Wonder 50

      Investigative Phenomenon: Waves 52

   Learn 62

   Share 82

Concept 4.3 Reducing Earthquake Impacts

Concept Overview 94

    Wonder. 96

        Investigative Phenomenon: The Aftermath of Earthquake 98

    Learn. 104

    Share 132

Unit Wrap-Up

Unit Project: Designing a Safe Bridge 142

Grade 4 Resources

Bubble Map. R3

Safety in the Science Classroom. R4

Vocabulary Flashcards R7

Glossary. R19

Index. R50

UNIT 4 | Earthquakes

Dear Parent/Guardian,

This year, your student will be using Science Techbook™, a comprehensive science program developed by the educators and designers at Discovery Education and written to the California Next Generation Science Standards (NGSS). The California NGSS expect students to act and think like scientists and engineers, to ask questions about the world around them, and to solve real-world problems through the application of critical thinking across the domains of science (Life Science, Earth and Space Science, Physical Science).

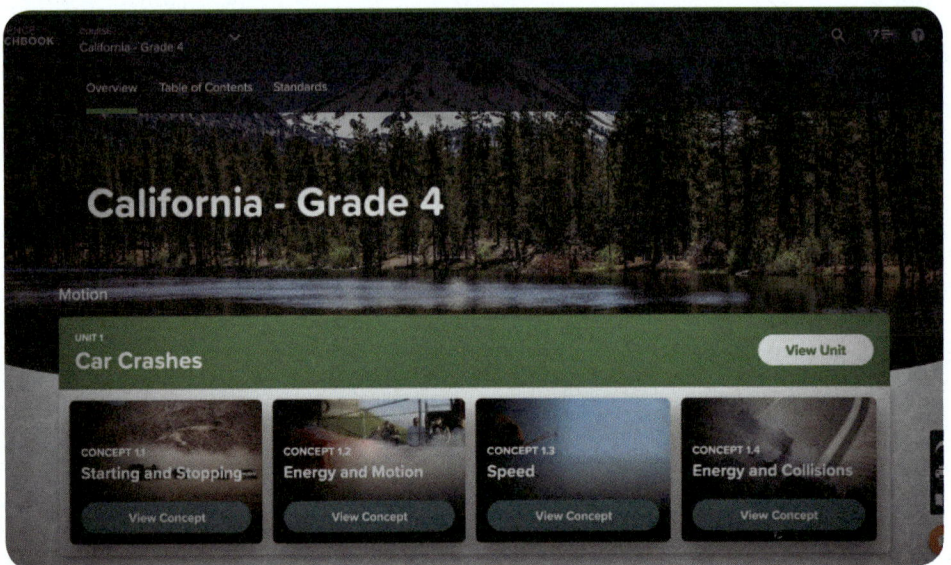

Science Techbook is an innovative program that helps your student master key scientific concepts. Students engage with interactive science materials to analyze and interpret data, think critically, solve problems, and make connections across science disciplines. Science Techbook includes dynamic content, videos, digital tools, Hands-On Activities and labs, and game-like activities that inspire and motivate scientific learning and curiosity.

You and your child can access the resource by signing in to www.discoveryeducation.com. You can view your child's progress in the course by selecting the Assignment button.

© Discovery Education | www.discoveryeducation.com

Science Techbook is divided into units, and each unit is divided into concepts. Each concept has three sections: Wonder, Learn, and Share.

Units and Concepts Students begin to consider the connections across fields of science to understand, analyze, and describe real-world phenomena.

Wonder Students activate their prior knowledge of a concept's essential ideas and begin making connections to a real-world phenomenon and the **Can You Explain?** question.

Learn Students dive deeper into how real-world science phenomenon works through critical reading of the Core Interactive Text. Students also build their learning through Hands-On Activities and interactives focused on the learning goals.

Share Students share their learning with their teacher and classmates using evidence they have gathered and analyzed during Learn. Students connect their learning with STEM careers and problem-solving skills.

Within this Student Edition, you'll find QR codes and quick codes that take you and your student to a corresponding section of Science Techbook online. To use the QR codes, you'll need to download a free QR reader. Readers are available for phones, tablets, laptops, desktops, and other devices. Most use the device's camera, but there are some that scan documents that are on your screen.

For resources in California Science Techbook, you'll need to sign in with your student's username and password the first time you access a QR code. After that, you won't need to sign in again, unless you log out or remain inactive for too long.

We encourage you to support your student in using the print and online interactive materials in Science Techbook, on any device. Together, may you and your student enjoy a fantastic year of science!

Sincerely,

The Discovery Education Science Team

Unit 4
Earthquakes

Get Started

A Bridge in an Earthquake

Earthquakes shake buildings and other structures. How these structures behave depends on the strength of the earthquake and the design of the structures.

Quick Code: ca4755s

A Bridge in an Earthquake

Think About It

Look at the picture. **Think** about the following questions.

- How have earthquakes shaped California's history?
- How can we describe the amount of shaking in earthquakes?
- How can we minimize the damage from earthquakes?

Bridge

Unit 4: Earthquakes

Unit Project Preview

Solve Problems Like a Scientist

Quick Code: ca4756s

Unit Project: Designing a Safe Bridge

In this project, you will use what you know about earthquake impacts and about engineering and technology to design an earthquake-proof bridge.

Bay Bridge

SEP Constructing Explanations and Designing Solutions
CCC Structure and Function

Ask Questions About the Problem

You are going to design an earthquake-proof bridge using what you know about why earthquakes occur and how they cause damage. **Write** some questions you can ask to learn more about the problem. As you learn about forces in this unit, **write** down the answers to your questions.

Unit 4: Earthquakes

CONCEPT 4.1

Earthquake Causes and Impacts

Student Objectives

By the end of this lesson:

- ☐ I can develop models of waves that describe how waves cause objects to move.
- ☐ I can explain how amplitude and wavelength affect the movement of seismic waves through Earth.
- ☐ I can construct a seismometer and plan and carry out tests to improve it.
- ☐ I can compare tsunami warning and evacuation systems based on how well they protect people.

Key Vocabulary

- ☐ air
- ☐ amplitude
- ☐ energy
- ☐ light
- ☐ seismic wave
- ☐ sound
- ☐ sound wave
- ☐ tsunami
- ☐ wave
- ☐ wavelength

Quick Code: ca4773s

Concept 4.2: Earthquake Waves

Activity 1
Can You Explain?

How does an earthquake deep in Earth cause damage to buildings situated many kilometers away?

Quick Code: ca4775s

4.2 | Wonder

How does an earthquake deep in Earth cause damage to buildings situated many kilometers away?

Activity 2

Ask Questions Like a Scientist

Quick Code: ca4776s

Waves

Look at the image and **answer** the questions. Then, **complete** the activity that follows.

Let's Investigate Waves

 CCC Energy and Matter

What do you think caused the waves in this picture?

_____

_____

Where else have you seen waves?

_____

_____

Can you think of other types of waves aside from water waves?

_____

_____

With your group, **share** additional examples of waves.

For each of the examples of waves you identified, how do those waves involve energy?

_____

_____

Now, **categorize** your explanations of energy into one of three categories: energy added used to create the motion of the wave, energy transferred by the wave, or energy within the wave. **Use** a list or chart to display your data.

Concept 4.2: Earthquake Waves

4.2 | Wonder

How does an earthquake deep in Earth cause damage to buildings situated many kilometers away?

Drop your object into the paper cup of water and **answer** the questions that follow.

Did the water have energy before you dropped the object?

_____

Where does the energy to make the waves come from?

_____

_____

What energy was transferred?

_____

_____

Earthquake waves move in a similar way to what you observed. Based on this statement, what questions do you have related to the energy of an earthquake?

_____

_____

_____

Activity 3
Analyze Like a Scientist

Quick Code: ca4777s

What Is a Wave?

Read the text. Then, **complete** the activity that follows.

What Is a Wave?

Think about how **waves** are made. Making waves requires **energy**. In the case of the beach, that energy comes from the wind. In the swimming pool or bathtub, the waves are made by people splashing around. Waves move energy from one place to another.

Thinking about waves helps us understand earthquakes. You know that earthquakes are made when energy is released at a fault. That energy travels away from the fault. It travels through the ground as waves.

Demonstrate with your hands or body how you think waves move in a pool and from an earthquake.

Swimming in a Pool

SEP Developing and Using Models
SEP Obtaining, Evaluating, and Communicating Information

Concept 4.2: Earthquake Waves

An analogy describes the relationship or shared characteristic or feature between two or more objects. For example, "Nest is to a bird as beehive is to a bee." Create an analogy comparing how earthquake waves are similar to waves on the beach.

_____

_____

_____

_____

_____

Activity 4

Observe Like a Scientist

Shockwave

Quick Code: ca4778s

Watch the video and **look** for different types of waves and then **discuss** what you have observed.

Shockwave

Talk Together

With a partner, take turns talking about one type of wave from the video and the material through which the wave traveled. As your partner shares, listen and then restate what your partner shared. Then, suggest an additional detail to add to the description.

SEP Asking Questions and Defining Problems

Concept 4.2: Earthquake Waves | 57

4.2 | Wonder

How does an earthquake deep in Earth cause damage to buildings situated many kilometers away?

Activity 5
Evaluate Like a Scientist

Quick Code: ca4779s

What Do You Already Know About Earthquake Waves?

Wavelength and Amplitude

Analyze the three waveforms shown. Then, **complete** the statements comparing these waveforms. Finally, **complete** the follow-up activity.

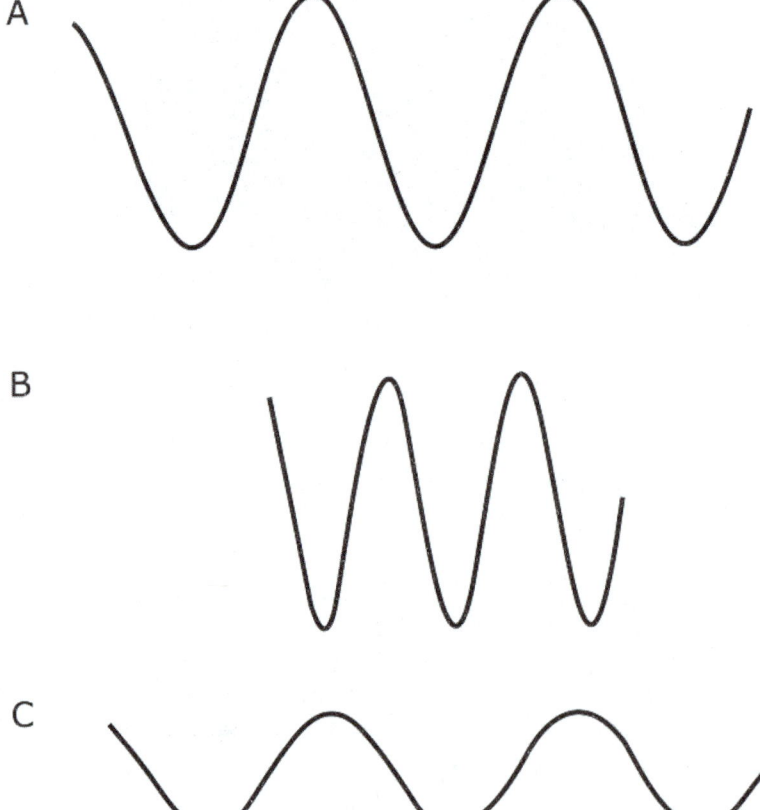

CCC Patterns

Choose the phrase from the list at the end of each sentence that makes the sentence correct. **Write** the phrase in the blank.

_____ has the shortest wavelength of the three waves. [Wave A, Wave B, Wave C]

Wave A has _____ Wave B. [a smaller amplitude than, a larger amplitude than, or the same amplitude as]

Wave C has the _____ amplitude of the three waves. [largest, smallest]

Wave C has the same wavelength as _____. [Wave A, Wave B]

Draw a wave similar but not identical to those shown.

Classify your group members' waves according to amplitude and wavelength. You can use rulers or meter sticks to **measure** and **compare** the waves.

Concept 4.2: Earthquake Waves | 59

4.2 | Wonder
How does an earthquake deep in Earth cause damage to buildings situated many kilometers away?

Analyze the Effects of Earthquakes

Look at the image and **answer** the question.

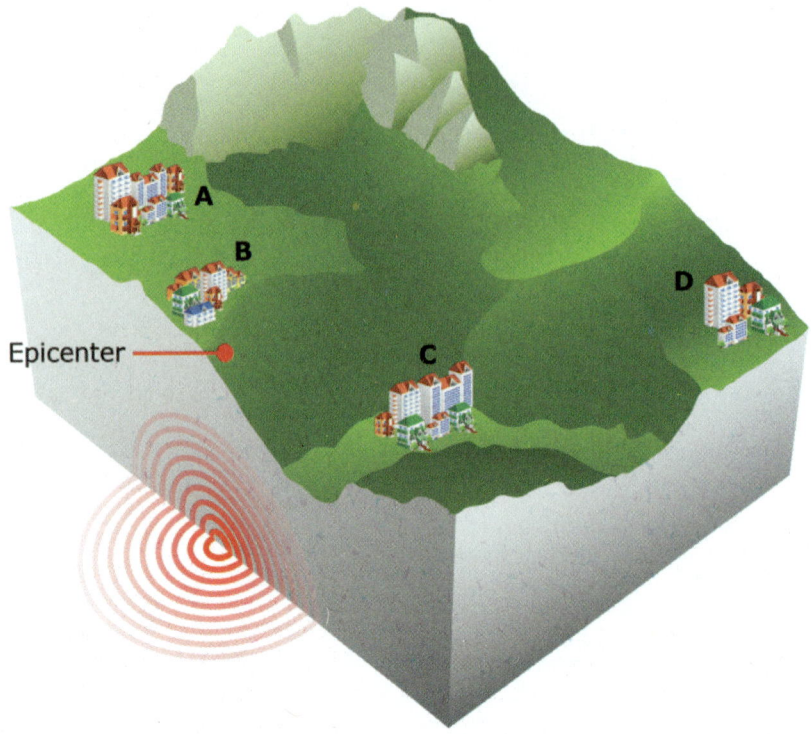

How will the earthquake affect the towns in the image? **Complete** the sentence with the letter of the two towns being described.

Town _____ would receive the largest shock, and

Town _____ would be the last to feel the earthquake.

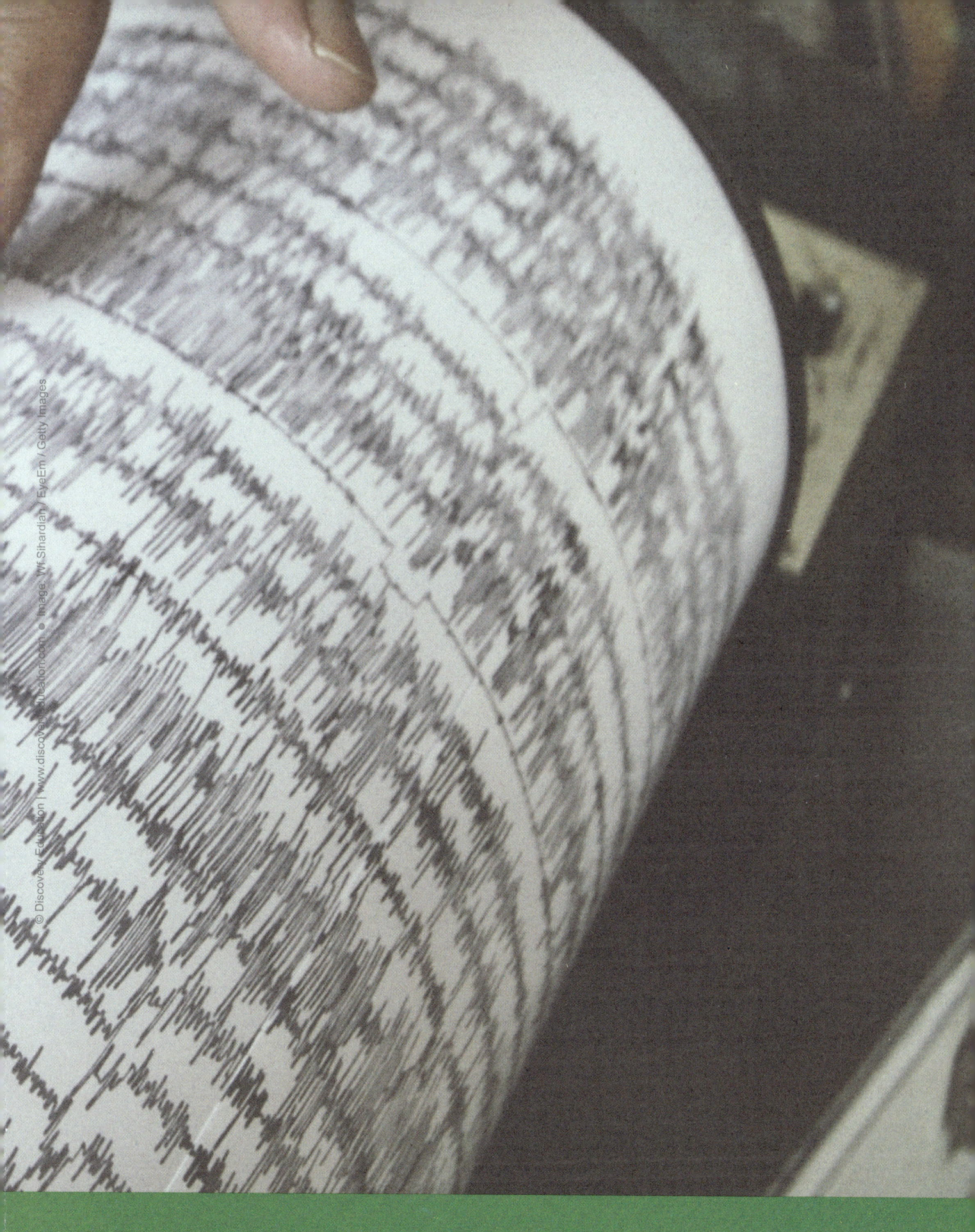

Concept 4.2: Earthquake Waves

4.2 | Learn
How does an earthquake deep in Earth cause damage to buildings situated many kilometers away?

What Are Waves, and How Do They Travel through a Medium?

Activity 6

Think Like a Scientist

Model Earthquake Waves

Quick Code: ca4780s

In this investigation, you will explore how earthquake waves are produced. You will also describe waves based on their characteristics.

What materials do you need? (per group)
- String

What Will You Do?

1. With your partner, make waves with the string.

2. You are going to examine the factors that affect the height and length of waves. Write the questions you want to answer in your investigation sheet.

3. With your partner, formulate a hypothesis about the factors that affect the height and length of waves.

SEP Developing and Using Models

4. Plan an investigation to test your hypothesis.
5. Write your plan in your investigation sheet.
6. Carry out your plan and write your observations in your investigation sheet.
7. Write your conclusions based on your observations in your investigation sheet.

Think About the Activity

Compare the height of your wave with the height of the wave of another pair of students. Which wave would cause more shaking? How do you know?

_____
_____
_____

How does the movement of your hands and the string relate to earthquake waves?

_____
_____
_____

Earthquakes that occur under Earth's surface are felt as shaking on Earth's surface. Describe how you would model this.

_____
_____
_____
_____

Concept 4.2: Earthquake Waves | 63

4.2 | Learn
How does an earthquake deep in Earth cause damage to buildings situated many kilometers away?

Activity 7
Observe Like a Scientist

Quick Code: ca4781s

Mediums and Waves

Watch the video and **listen** for the definition of *medium*.

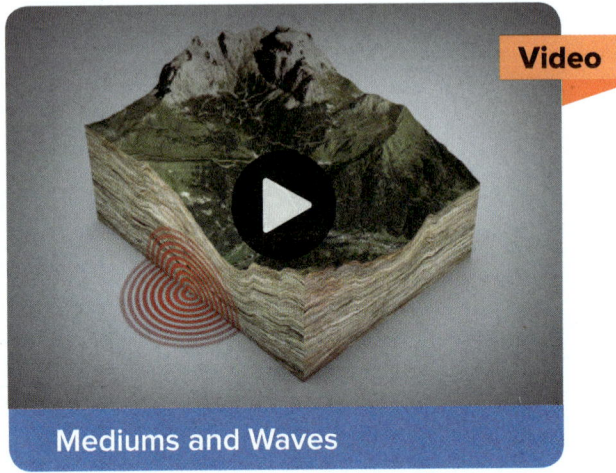

Mediums and Waves

Talk Together

Now, talk together about the definition of *medium* based on evidence from the video.

Activity 8

Analyze Like a Scientist

Quick Code: ca4782s

Moving Waves

Read the text and **look** at the image. **Highlight** any information you do not understand.

Moving Waves

Waves are everywhere. You see **light** waves with your eyes. You hear **sound waves** with your ears. You feel water waves against your skin. Waves are energy in motion. Energy moves from one place to another through a wave. A medium is the material through which a wave travels. It is made up of particles. Particles can be invisible. The **air** around us is made up of invisible particles. Particles can also make up solids. The desk or table at which you are sitting is made up of particles. When waves move through a medium, the energy from the wave is passed from one particle to the next. In this way, energy travels through the medium.

There are many different types of waves. They include water waves, sound waves, and earthquake waves. Earthquake waves, or **seismic waves**, are waves that travel through the ground.

When waves travel through a medium, such as air, water, or the ground, particles in the medium transmit the energy they carry, but the particles do not travel with the waves. For example, ocean waves move forward, but the

SEP	Developing and Using Models
SEP	Obtaining, Evaluating, and Communicating Information
CCC	Energy and Matter

Concept 4.2: Earthquake Waves | 65

water (and its particles) only moves up and down. What do you think happens to the ground when a wave passes through it?

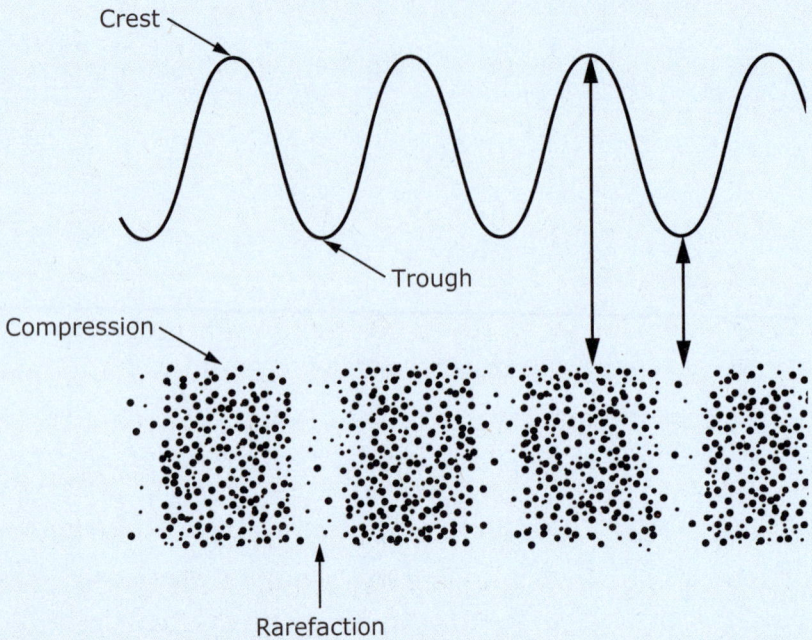

Now, **read** the text again. Then, **create** your own model of a wave.

How Are Seismic Waves Formed?

Activity 9
Analyze Like a Scientist

Amplitude and Wavelength

Read the text and **watch** the video. Then, **complete** the activity that follows.

Quick Code: ca4783s

Amplitude and Wavelength

Waves have two important properties: **amplitude** and **wavelength**. Amplitude is the height of a wave. In the ocean, waves can have different heights. Surfers love the taller waves! The amplitude of a wave is determined by its energy. The more energy a wave has, the greater its amplitude. Think about ocean waves. The taller the wave, the more energy it is carrying.

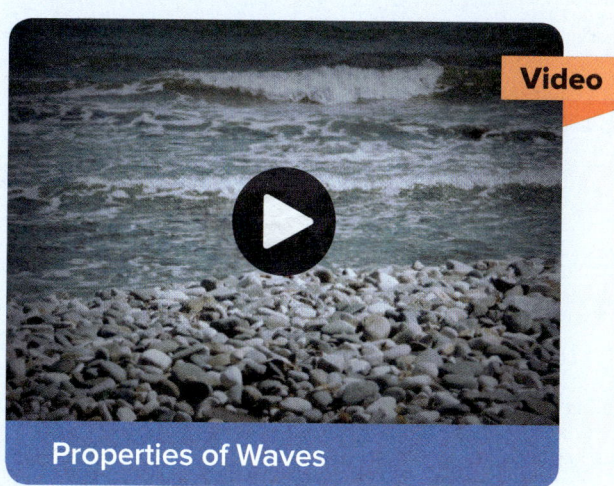

Properties of Waves

SEP Obtaining, Evaluating, and Communicating Information
CCC Patterns
CCC Energy and Matter

Concept 4.2: Earthquake Waves | 67

Amplitude and Wavelength cont'd

You know this if you have ever been knocked over by a big wave. The same applies to a seismic wave—the greater its amplitude, the more energy it has and the more damage it can do.

A second property of waves is wavelength. Wavelength is the distance between two similar parts of a wave. For example, it may be the distance between the tops of two waves. It could also be the distance between the lowest points of two waves. The wavelength of a seismic wave determines how often the ground shakes. The shorter the wavelength, the more rapidly the ground shakes.

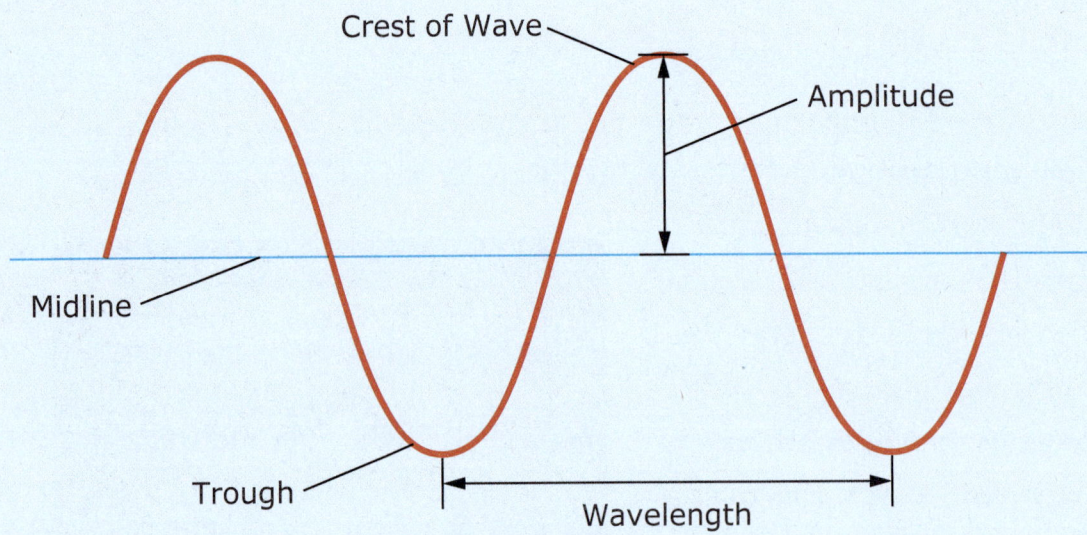

Both wavelength and amplitude affect the patterns that waves make. Waves with short wavelengths will occur in rapid patterns. Waves with longer wavelengths will be spaced out more. Waves with large amplitude will be much higher than waves with small amplitude. For example, ripples are waves with small amplitude. As a wave moves across a medium, its wavelength and amplitude can change. This means the pattern of waves can also change.

Trade wave models with a partner. **Create** another wave, based on your partner's wave model, with either larger or smaller magnitude.

4.2 | Learn

How does an earthquake deep in Earth cause damage to buildings situated many kilometers away?

How Do Scientists Discover and Study Seismic Waves?

Activity 10

Observe Like a Scientist

Recording Seismic Waves

Quick Code: ca4784s

Watch the animation and **note** the different parts of a seismogram. Next, **look** at the image. Then, **answer** the question that follows.

A Simple Seismometer

SEP Analyzing and Interpreting Data
CCC Patterns

List the different parts of a seismogram.

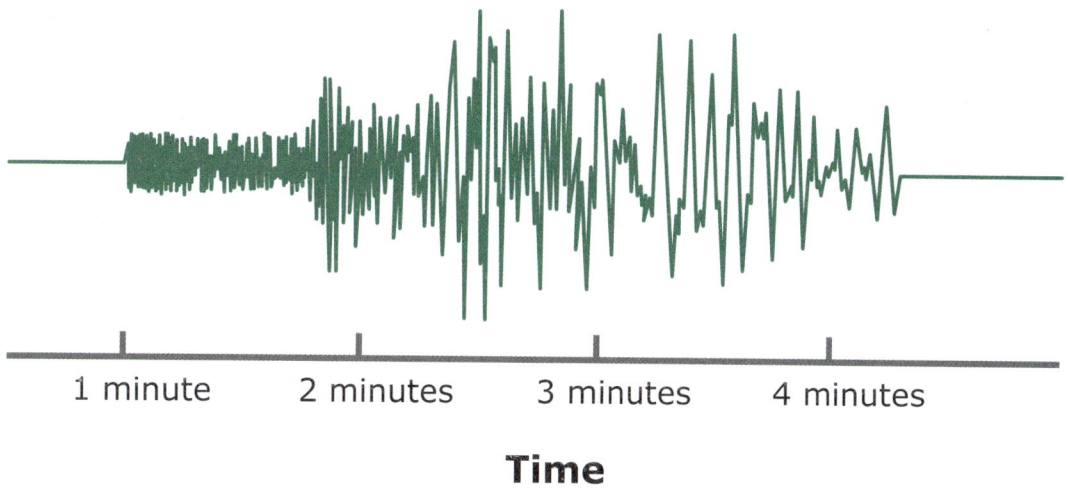

Time

What conclusions can you draw from the seismogram?

_____

_____

_____

_____

Concept 4.2: Earthquake Waves | **71**

4.2 | Learn
How does an earthquake deep in Earth cause damage to buildings situated many kilometers away?

Activity 11
Design Solutions Like a Scientist

Quick Code: ca4785s

Hands-On Engineering: Make a Seismometer

In this investigation, you will design a seismometer. You will then test your seismometer and make improvements to the design based on the tests you conducted.

Ask Questions About the Problem

What parts should your seismometer have?

_____

_____

How should these parts work together for the seismometer to work properly?

_____

_____

_____

Draw two possible designs based on the materials you have. **Label** the parts of your seismometer and describe the function of each part.

| SEP | Constructing Explanations and Designing Solutions |

72 |

What materials do you need? (per group)

- Shoebox or similarly sized and shaped cardboard box
- Scissors
- Metric ruler
- Paper clips
- Hole punch
- String
- Yarn
- 1-liter soda bottle
- Water
- Construction paper
- Markers
- Masking tape
- White paper
- Modeling clay
- Plastic cup, 9 oz
- Lids for cups

What Will You Do?

1. On a T-Chart, **list** the advantages and disadvantages of each design.
2. **Choose** the design you think is best based on the list you made.
3. **Construct** your seismometer.
4. **Discuss** how your design could be improved.
5. **Update** your drawing to include the improvements you decided on.

4.2 | Learn

How does an earthquake deep in Earth cause damage to buildings situated many kilometers away?

Think About the Activity

What needs did your design address and what were the measures of success? What were the limits on your project?

_____
_____
_____
_____
_____
_____
_____
_____

How many designs did you develop, and how well did your chosen design work? How did you test your design? What kind of problems did you run into when making your seismometer? How did you address these problems?

_____
_____
_____
_____
_____
_____
_____
_____

Compare your seismometer to that of another group. How are they the same? How are they different? What additional materials or features could you use to improve your design?

4.2 | Learn

How does an earthquake deep in Earth cause damage to buildings situated many kilometers away?

Activity 12
Observe Like a Scientist

Quick Code: ca4786s

Find the Temblor

Complete the interactive. **Notice** how P waves and S waves differ.

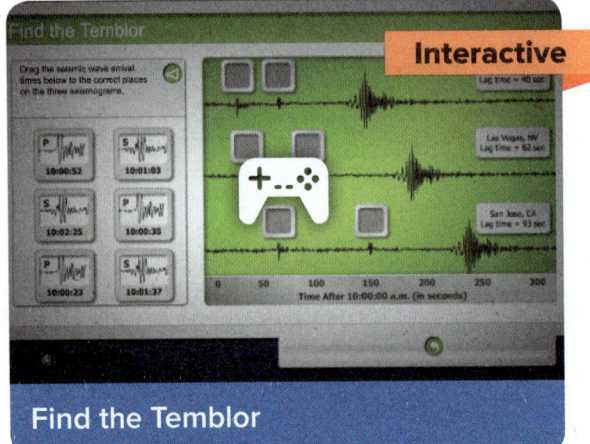

Find the Temblor

Talk Together

Now, talk together about patterns you see in P waves and S waves. Why does it take data from three stations to pinpoint the epicenter of an earthquake?

SEP Analyzing and Interpreting Data

76

Activity 13
Observe Like a Scientist

Waves in Action

Quick Code: ca4787s

Look at the image and **answer** the question that follows.

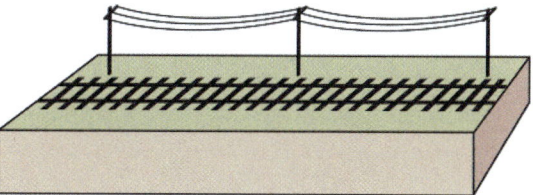

1. Appearance before an earthquake.

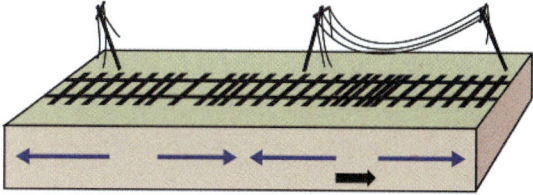

2. When an earthquake starts, the ground is stretched and squashed.

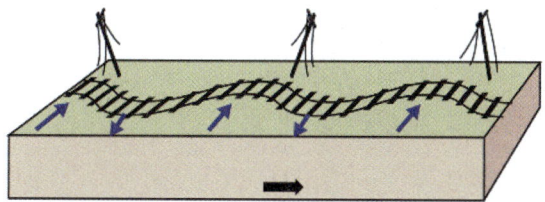

3. The next waves to arrive make the ground move from side to side.

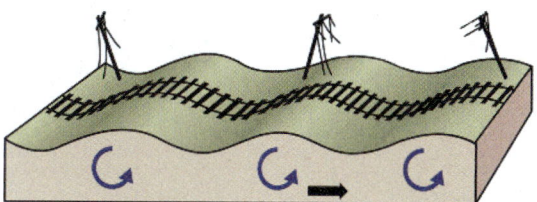

4. The last waves to arrive make the ground move up and down.

Which wave is represented in each image?

_____
_____
_____
_____

SEP Engaging in Argument from Evidence
CCC Cause and Effect

Concept 4.2: Earthquake Waves

4.2 | Learn
How does an earthquake deep in Earth cause damage to buildings situated many kilometers away?

How Do Earthquakes Cause Tsunamis?

Activity 14
Observe Like a Scientist

Tsunami

Watch the video and **complete** the cause-and-effect chart using examples from the video.

Quick Code: ca4788s

Tsunami

SEP Engaging in Argument from Evidence
CCC Cause and Effect

Cause	Event	Effect

Concept 4.2: Earthquake Waves | 79

4.2 | Learn — How does an earthquake deep in Earth cause damage to buildings situated many kilometers away?

Activity 15

Evaluate Like a Scientist

Quick Code: ca4789s

Tsunami Warning System

Tsunamis can move at speeds of 900 kilometers per hour. They can be 20 meters tall and can reach hundreds of meters inland. Imagine that you are in charge of developing a tsunami warning and evacuation system. Working with a partner, **create** an illustrated description of your plan on a sheet of poster paper. Your design should be labeled and should explain the purpose of each of the system components.

After viewing your classmates' presentations, how might you improve your design?

_____

_____

_____

_____

SEP Constructing Explanations and Designing Solutions

Concept 4.2: Earthquake Waves

4.2 | Share
How does an earthquake deep in Earth cause damage to buildings situated many kilometers away?

Activity 16
Record Evidence Like a Scientist

Quick Code: ca4790s

Waves

Now that you have learned about waves, **look** again at the image of waves. You first saw this in Wonder.

Let's Investigate Waves

Talk Together

How can you describe waves now? How is your explanation different from before?

SEP Constructing Explanations and Designing Solutions

Look at the Can You Explain? question. You first read this question at the beginning of the lesson.

> **Can You Explain?**
>
> How does an earthquake deep in Earth cause damage to buildings situated many kilometers away?

Now, you will use your new ideas about earthquake waves to answer a question.

1. **Choose** a question. You can use the Can You Explain? question or one of your own. You can also use one of the questions that you wrote at the beginning of the lesson.

   My Question

2. Then, use the graphic organizers on the next pages to help you answer the question.

Concept 4.2: Earthquake Waves | 83

4.2 | Share

How does an earthquake deep in Earth cause damage to buildings situated many kilometers away?

To plan your scientific explanation, first **write** your claim. Your claim is a one-sentence answer to the question you investigated. It answers: What can you conclude? It should not start with yes or no.

My claim:

Evidence must be sufficient: use enough evidence to support the claim. Next, **identify** two pieces of evidence that support your claim:

Evidence 1

Evidence 2

4.2 | Share

How does an earthquake deep in Earth cause damage to buildings situated many kilometers away?

Finally, **explain** your reasoning. Reasoning ties together the claim and the evidence. Reasoning shows how or why the data count as evidence to support the claim.

Evidence	Reasoning That Supports Claim

Now, **write** your scientific explanation.

An earthquake deep in Earth can cause damage to buildings situated many kilometers away because:

_____
_____
_____
_____
_____
_____
_____
_____
_____
_____
_____
_____
_____
_____
_____
_____
_____

STEM in Action

Activity 17
Analyze Like a Scientist

Quick Code: ca4791s

Careers: Seismologists

Read the text and **watch** the video. Then, **answer** the questions that follow.

Careers: Seismologists

Seismologists are scientists who research earthquakes and their causes. They do this research by studying earthquake waves. The data collected from the waves tell the seismologists the strength of the earthquake and where it happened.

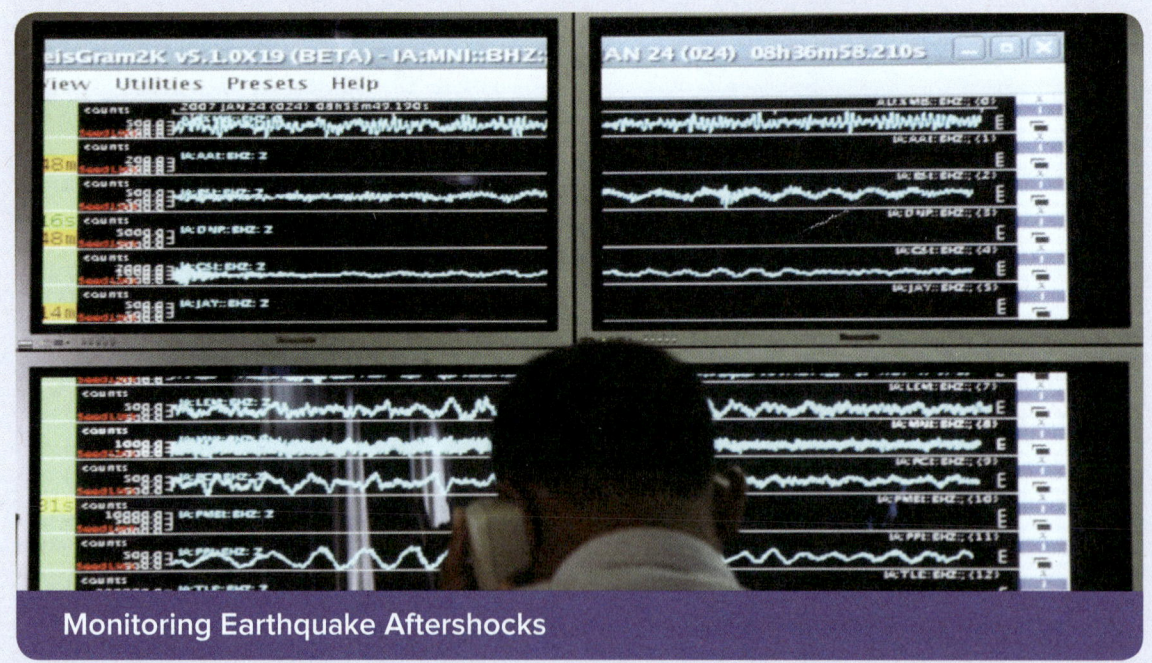

Monitoring Earthquake Aftershocks

Seismologists work in offices, in the field, and in universities. In the office, they use computers to model earthquakes. The models help them to understand how earthquakes happen and how earthquakes could affect certain areas. In the field, they install instruments that can help them learn about earthquakes. In universities, they teach students seismology and perform seismological research. Seismology is the study of earthquakes.

Seismologist

Seismologists also work with cities and towns in areas where earthquakes may occur. They help the communities design buildings and homes that are strong enough to withstand earthquakes. By doing this, seismologists help save lives and limit damage after an earthquake.

Seismologists use several instruments to help them do their job. They use a seismograph to find the epicenter and the strength of an earthquake. Seismologists also use the Global Positioning System (GPS). GPS analyzes data from satellites to tell people their exact location. Seismologists use GPS to monitor and measure the movement of rocks that make up Earth's surface. GPS has also helped scientists understand how rocks move under Earth's surface.

Can seismologists accurately predict earthquakes? Even though seismologists around the world have researched earthquakes for a long time, they still cannot predict exactly when and where an earthquake will happen. Someday, better instruments may help scientists to predict earthquakes more accurately.

Concept 4.2: Earthquake Waves

Think Like a Seismologist

This map shows the occurrence of earthquakes in the United States from January 4 to January 10, 2018. The numbers represent the number of earthquakes at that location during that time.

Look at the map.

What conclusion can you draw from the map?

_____
_____
_____
_____

SEP Obtaining, Evaluating, and Communicating Information

Scientists and Communities

Suppose you are the mayor of your town. The city approved a bond to fund research to make sure that the community's buildings are earthquake safe. Which scientist and other people will you work with to do this job? Describe how you will work with the scientist and other people, and include your goal in working with them.

_____

_____

_____

_____

_____

_____

_____

Concept 4.2: Earthquake Waves

4.2 | Share
How does an earthquake deep in Earth cause damage to buildings situated many kilometers away?

Activity 18
Evaluate Like a Scientist

Quick Code: ca4792s

Review: Earthquake Waves

Think about what you have read and seen in this lesson. **Write** down some core ideas you have learned. **Review** your notes with a partner. Your teacher may also have you take a practice test.

_____
_____
_____
_____
_____
_____

 **Talk Together**

Think about what you saw in Get Started. Use your new ideas about earthquake waves to discuss how engineers can design bridges for areas with earthquakes.

SEP Obtaining, Evaluating, and Communicating Information

CONCEPT 4.3

Reducing Earthquake Impacts

Student Objectives

By the end of this lesson:

- ☐ I can obtain and combine information to generate and compare multiple earthquake preparedness plans based on how well they reduce the impact of earthquakes on humans.

- ☐ I can plan and carry out tests of earthquake-proof houses to identify how they can be improved.

Key Vocabulary

- ☐ engineer
- ☐ model
- ☐ seismic
- ☐ system

Quick Code: ca4793s

Concept 4.3: Reducing Earthquake Impacts

Activity 1
Can You Explain?

What should I do to keep myself safe during and after an earthquake?

_____

_____

_____

_____

_____

_____

_____

_____

_____

_____

Quick Code: ca4795s

Concept 4.3: Reducing Earthquake Impacts | 97

4.3 | Wonder
What should I do to keep myself safe during and after an earthquake?

Activity 2
Ask Questions Like a Scientist

The Aftermath of an Earthquake

Quick Code: ca4796s

Watch the video and **complete** the activity that follows.

Video

Let's Investigate the Aftermath of an Earthquake

SEP Asking Questions and Defining Problems

What questions do you have about the impact of earthquakes?

_____

Work with your group to **complete** the T-chart.

Topic: _____

Type of Impact	Suggestions of how impact can be reduced

Concept 4.3: Reducing Earthquake Impacts

Activity 3
Analyze Like a Scientist

Quick Code: ca4797s

Loss of Life from Earthquakes

Read the text and **look** at the pictures. **Answer** the question that follows the text.

Loss of Life from Earthquakes

In Kobe, Japan, thousands of people died when an earthquake struck their city. However, in Japan, people are prepared for earthquakes. The impact, although terrible, could have been much worse.

Earthquake in Kobe, Japan

SEP Constructing Explanations and Designing Solutions
CCC Cause and Effect

In a poorer country, Haiti, an earthquake hit Port-au-Prince. In this city of over 2 million people, more than 230,000 people died. A similar earthquake in Oakland, California, killed 63 people.

Earthquake in Port-au-Prince

Oakland Earthquake

What evidence describes why there was such a difference in the loss of life at each location for the earthquakes?

_____

_____

_____

_____

_____

_____

Concept 4.3: Reducing Earthquake Impacts

4.3 | Wonder — What should I do to keep myself safe during and after an earthquake?

Activity 4

Evaluate Like a Scientist

Quick Code: ca4798s

What Do You Already Know About Reducing Earthquake Impacts?

Who Needs to Prepare?

Read the statements. **Circle** all of the true statements.

If you have not experienced an earthquake, you should prepare for earthquakes.

If you have experienced an earthquake, you do not need to prepare further for earthquakes.

If your area does not expect earthquakes, you do not need to prepare for earthquakes.

If your area is at risk for earthquakes, you should prepare for earthquakes.

Earthquake Precautions

Read each statement. **Decide** if it applies to earthquakes. If it does not, **write** "No." If it does, **write** whether it applies "before" or "during" an earthquake.

Move your furniture to the upper levels of the house if possible. _____

Attach heavy and tall furniture to the walls. _____

Try to move to the innermost part of the house. _____

Avoid windows, mirrors, and things hanging on the walls. _____

If driving, get out of the vehicle and lie down in the lowest area possible. _____

Create an emergency plan including everyone in the house. _____

Stop where you are, get low, and cover your head and neck. _____

Concept 4.3: Reducing Earthquake Impacts | 103

4.3 | Learn What should I do to keep myself safe during and after an earthquake?

How Can I Prepare for an Earthquake?

Activity 5

Think Like a Scientist

Quick Code: ca4799s

Earthquake Safety Zones

In this investigation, you will **create** a map of your home showing safety zones during an earthquake.

What Will You Do?

Draw a map of your house and **mark** safe and unsafe areas. **Include** at least three safe areas.

Create a map of safe areas near your classroom.

SEP Developing and Using Models

SEP Engaging in Argument from Evidence

Concept 4.3: Reducing Earthquake Impacts

4.3 | Learn — What should I do to keep myself safe during and after an earthquake?

Activity 6
Observe Like a Scientist

Quick Code: ca4800s

Safety Tips to Use during Earthquakes

Watch the video and **look** for tips that you can add to your safety map.

Safety Tips to Use during Earthquakes

Talk Together

Now, talk together about the tips you have added to your safety map.

Activity 7
Analyze Like a Scientist

Earthquake Preparation

Read the text and look at the pictures. Your teacher will provide you with sticky notes. **Use** these notes to write a modification to a solution, a new idea, or an answer to one of the questions given in the article.

Quick Code: ca4801s

Earthquake Preparation

Earthquakes can occur anywhere at any time. Many people are caught unprepared and do not know what to do when an earthquake strikes. Preparation can help prevent injury, deaths, and property damage.

Earthquake preparation requires smart planning. We do not have a warning **system** to tell us when an earthquake will occur, so earthquake preparation should be a high priority for all people. Being prepared for a disaster can reduce its effects and make everyone safer.

A car is crushed beneath the rubble of a ruined house after an earthquake in Chile. How can people prepare for earthquakes?

SEP Constructing Explanations and Designing Solutions
SEP Obtaining, Evaluating, and Communicating Information
CCC Cause and Effect

Concept 4.3: Reducing Earthquake Impacts

Earthquake Preparation *cont'd*

Reporters and scientists take notes at a seismograph station where earthquake activity is monitored. Why are earthquakes so dangerous?

Shaking Ground

The shaking of the ground caused by **seismic** waves is not by itself hazardous. If you are in an open area, such as field, you should be safe. The biggest hazard during an earthquake is collapsing buildings. Once the earthquake is over, there are other hazards. Fires often occur after earthquakes as gas and power lines break. Water pipes can also break, making firefighting difficult.

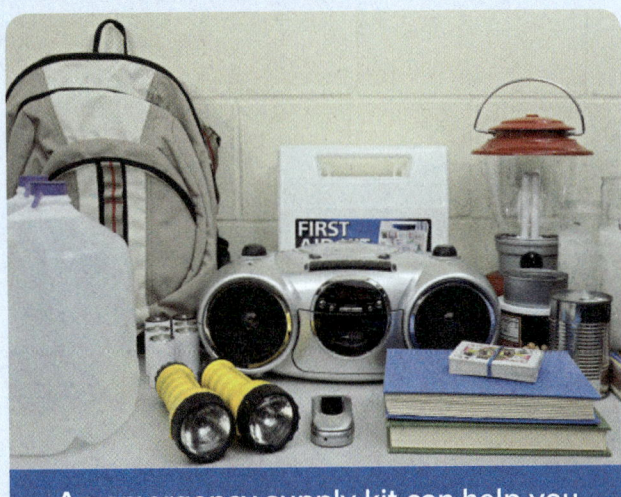

An emergency supply kit can help you survive after an earthquake. How else can you prepare for an earthquake?

Aftershocks

Many earthquakes are followed by other earthquakes called aftershocks. These can be very dangerous because some people relax their guard after the main earthquake and do not continue to follow the correct precautions. It is important to be prepared for the earthquake and its aftermath.

Tsunami

If you live near the coast, a tsunami triggered by the earthquake could also occur. Because earthquakes under the ocean cause tsunamis, there may be a warning before they reach the shore. Tsunami warning centers determine the location and size of an earthquake to determine the potential of tsunami generation and predict tsunami wave arrival times and locations. They can then give warnings to the population in the hazard zone. Earthquakes that generate tsunamis usually occur close enough to the coast that waves arrive within one or two hours.

A person on or near the coast following an earthquake should take important steps to avoid being harmed in the event of a tsunami. These steps include locating tsunami evacuation signs, listening carefully to all announcements, and heading for and remaining on higher ground until the all-clear is received. If you live in a coastal area prone to earthquakes, it is good to have a tsunami evacuation plan.

If a tsunami warning is given, follow the evacuation route recommended. If there are no signs, why should you go to high ground?

Concept 4.3: Reducing Earthquake Impacts

Activity 8
Analyze Like a Scientist

Preparing for an Earthquake

Quick Code: ca4802s

Read the text. Then, **answer** the questions that follow.

Preparing for an Earthquake

Earthquakes can happen at any time, in any weather. Seismologists can know which areas are more likely to experience earthquakes, but they cannot know exactly when one will happen. Preparedness is key when it comes to earthquakes. You can prepare for a major earthquake by Planning, Packing, Protecting, and Practicing.

Plan

Every family should have an earthquake emergency plan. The plan should make it clear what each family member should do when an earthquake hits.

During the Earthquake

Every family member should stop what they are doing, get on the ground, crawl to a nearby safe zone, and cover their neck and head. Stop, drop, and hold on!

This part of the plan should detail where the safety zones in your home are.

SEP Constructing Explanations and Designing Solutions
SEP Obtaining, Evaluating, and Communicating Information

Immediately After the Earthquake

All family members should cover their nose and mouth with cloth. This will help keep dust out of your lungs. Everyone should move carefully through the home: structure; things can become less stable after an earthquake. Family members should find their buddy and go to the meeting place. If a family member is trapped, they should move as little as possible and make some type of noise. Tapping on a hard object like a pipe or wall is best. Shouting should be a last resort.

This part of the plan should assign each family member—including pets—a buddy. It should also name a meeting place, like the driveway or a street corner.

A While After the Earthquake

An adult should check for small fires and put out any they find. If water or gas is leaking, an adult should shut off the utilities. If there are any injured people, they should be helped. If a person is seriously injured, they should not be moved unless it is dangerous to stay there. The family should locate their emergency supplies and monitor local news. The news on radio, television, social media, or text alerts can give information and instructions. The family should check in with an out-of-area friend or relative who can let others know that the family is safe. If the family is leaving the home, they should leave a message saying where they are headed.

This part of the plan should specify where and how to shut off utilities like water and gas. It should also indicate where the emergency supplies are located. An out-of-area friend or relative should be chosen, and a simple phone tree should be made.

Preparing for an Earthquake *cont'd*

The supplies in this emergency pack have been spread out on a table.

- Portable radio
- Batteries
- Blankets for each person
- Clothes for each person
- Shoes for each person
- Cash
- Medication
- An alternate cooking source

Preparing for an Earthquake

For larger families, it may be easier for each person to create his or her own kit. The kit or kits should be kept in a safe and easy-to-reach place.

It is also smart to keep a flashlight and shoes or hard-soled slippers by your bed. This will make it easier to move around in dark or debris.

Protect

You can protect your home from an earthquake by eliminating hazards. Move beds out from under windows. Make sure things hanging off walls or ceilings are secure. Anchor water heaters, furnaces, and tall, heavy furniture to the walls. Do not hang pictures on the wall behind your bed. Move heavy items to lower shelves or cabinets. Store hazardous or flammable liquids and breakables in low shelves or secured cabinets.

Practice

Every family member should know the earthquake emergency plan. The family should practice "stop, drop, and hold on" and gather in the meeting place. The more you practice your plan, the easier it will be to do in an actual emergency.

How can you make sure that your family is prepared for a major earthquake?

_____
_____
_____
_____
_____

How do we know if a preparation plan will work?

_____
_____
_____
_____
_____

Which of these preparation tips would work for a different disaster, like a hurricane? Which would not work, and why?

_____
_____
_____
_____
_____

4.3 | Learn — What should I do to keep myself safe during and after an earthquake?

Activity 9
Observe Like a Scientist

Making an Earthquake Kit

Quick Code: ca4803s

Watch the video. **Look** for and list the materials that the family has stored in preparation for an earthquake.

Making an Earthquake Kit

Talk Together

Now, talk together about additional items you think would be a good idea to add to an earthquake preparedness kit.

Activity 10
Design Solutions Like a Scientist

Quick Code: ca4804s

Hands-On Engineering: My Earthquake Preparedness Plan

In this investigation, you will **create** an earthquake preparedness plan for when your family is at home.

Ask Questions About the Problem

What can your family do to prepare for an earthquake?

_____
_____
_____
_____
_____
_____
_____
_____
_____
_____
_____
_____

SEP Constructing Explanations and Designing Solutions
SEP Developing and Using Models

Concept 4.3: Reducing Earthquake Impacts | 115

4.3 | Learn What should I do to keep myself safe during and after an earthquake?

What materials do you need? (per group)

- Earthquake Preparedness Plan sheet
- Pencils

What Will You Do?

1. **Decide** what the problem is that you are trying to solve.
   _____
   _____

2. **Write** a plan for what your family will do to prepare your furniture and home for an emergency.
   _____
   _____
   _____

3. **Draw** the supplies you will put into your emergency pack.

What should all family members do during the quake?

_____

Where are three of the safety zones in your home?

_____

_____

_____

What should all family members do as soon as the quake is over?

_____

_____

Who is buddies with whom in your family? Include any pets.

_____

_____

_____

Draw a map that shows your house, your in-neighborhood meeting place, and your out-of-neighborhood meeting place. Both meeting places should be a place that all family members know.

Concept 4.3: Reducing Earthquake Impacts

Activity 11
Analyze Like a Scientist

Quick Code: ca4805s

Surviving an Earthquake

Read the text. Then, **think** about and prepare to discuss the questions that follow.

Surviving an Earthquake

Unlike other large natural disasters, earthquakes are completely unpredictable. On what appears to be a perfectly normal day, the ground suddenly begins to shake. Areas close to the source of the quake receive no warning. Even people farther away may have only a few seconds to prepare. Still, there are actions that can be taken both before and during an earthquake that will limit damage and injury.

Baseball fans were not prepared for the earthquake that interrupted the 1989 World Series.

SEP Engaging in Argument from Evidence

Preparing for Earthquakes

Individuals, **engineers**, and national and local governments each have their own roles to play in preparing for earthquakes. Individuals can store emergency supplies and practice survival skills. Engineers can build quake-resistant structures and upgrade existing ones. Governments can enact laws requiring people to follow best practices for building quake-resistant structures. Governments can also create and fund disaster-response plans. Individuals can greatly improve their chances of surviving an earthquake by creating emergency supply kits. These kits should be stored in safe places. When choosing the items to place in an emergency kit, people should keep in mind that earthquakes can disrupt water, gas, and electrical lines. Items that require any of these utilities should not be included. The most important items are listed below:

- Enough food and water to last three days (one gallon of water per person per day); no food that requires cooking
- Manual can opener
- First aid kit
- Hand-crank or battery-powered radio
- Battery-powered flashlights
- Extra batteries
- Whistle to signal for help
- Tools needed to turn off utilities

Surviving an Earthquake *cont'd*

An emergency kit for earthquakes should include drinking water, first aid materials, battery-powered equipment, and other supplies shown here.

Damage to household items during a quake can be reduced by giving some thought to how materials are stored. Heavy items should be stored on lower shelves. Bookcases and other tall furniture must be securely attached to the wall. The water heater should be strapped to the wall as well. Individuals need to learn how to turn off gas, electricity, and water lines in their home. For more details on earthquake preparedness, people should visit the website of the Federal Emergency Management Agency (FEMA). The FEMA website also provides detailed instructions for how to protect oneself during and after a quake.

How to Prevent Damage

A list of statistics for major earthquakes shows that quakes of equal strength have caused very different amounts of damage and injuries. Most likely, an earthquake that happens directly beneath a city will be more disastrous than one that takes place far from population centers. Equally important is how the buildings that experience an earthquake are constructed. People who work in the field of earthquake engineering are concerned with why structures collapse during quakes and how to prevent this from happening.

Engineers have learned a lot from shake table experiments. The procedure is simply to shake a **model** of a proposed structure and observe the damage. These tests can be very expensive, but the data they provide can help to prevent even greater losses to the actual structure following an earthquake.

One basic strategy for making a structure is to build in mechanisms, called dampers, which absorb shock. Another technique is to isolate the structure from its base so that the base can move independently of the upper part. This can be as simple as resting a building on giant ball bearings.

Certain building materials should be avoided. The mortar between bricks and stones tends to dry out and crumble when it becomes very old. In the 1989 Loma Prieta earthquake in California, many of the buildings that collapsed were old, brick-and-mortar structures. The ancient Incan civilization of Peru may have known this because they built using carefully fitted stones without mortar. These stone buildings have survived many earthquakes.

The Taipei 101 skyscraper is equipped with a seismic damper to absorb shocks from seismic waves.

Concept 4.3: Reducing Earthquake Impacts

Surviving an Earthquake *cont'd*

The Inca built stone walls without mortar that have survived earthquakes for hundreds of years.

Earthquake engineers also make seismic retrofits to existing buildings. These strategies include isolating the building's base and adding bracing materials to the structure.

In states prone to earthquakes, building codes have been enacted that set standards for new construction and require retrofits of some existing structures. Another role of governments is to create plans for the aftermath of a large earthquake. Such plans ensure that firefighters, police officers, and other government employees all work together smoothly. Other parts of the plan will designate schools and other public buildings as evacuation centers and will map out evacuation routes. Many communities post their disaster plans on the websites of the relevant county or city government.

What do you think are the three most important things to pack in your emergency supplies?

_____
_____
_____
_____

Why do you think those three things are most important?

_____
_____
_____
_____

Why is it important to prepare before an earthquake happens?

_____
_____
_____
_____

Compare your plan with another group's plan. What is the same? What is different?

_____
_____
_____
_____

4.3 | Learn — What should I do to keep myself safe during and after an earthquake?

How Can We Make Buildings to Better Withstand Earthquakes?

Activity 12

Observe Like a Scientist

San Francisco's Skyscraper Design

Watch the video. **Look** for ways that many skyscrapers in San Francisco have been constructed to make them resistant to earthquakes.

Quick Code: ca4806s

San Francisco's Skyscraper Design

Talk Together

Now, talk together about how the different design features help make the building safe.

SEP Obtaining, Evaluating, and Communicating Information

Activity 13
Analyze Like a Scientist

Quick Code: ca4807s

Earthquake-Resistant Buildings

Read the text. As you read, **underline** sections of the text that will help you design an earthquake-resistant structure in the Hands-On Engineering activity.

Earthquake-Resistant Buildings

Imagine you are an engineer studying how earthquakes affect buildings. How could you go about your work? You could visit places where earthquakes had occurred and see what had happened to the buildings in the area. You could live in an area where earthquakes are common and wait for one to happen—but that could involve a lot of waiting!

Alternatively, you could model an earthquake. Engineers model earthquakes in a number of different ways. Sometimes, they use computer models. Other times they create special tables, called shake tables, which simulate the movement of the earth and test miniature versions of their designs. On a few occasions, they even shake entire buildings. From this work, they have learned a lot about the construction of earthquake-resistant buildings.

SEP Obtaining, Evaluating, and Communicating Information
CCC Structure and Function

Earthquake-Resistant Buildings *cont'd*

Design Features

If collapsing buildings are the main cause of damage, injury, and death, then one safety solution would be to make them earthquake-proof. Engineers have learned how to build earthquake-resistant buildings that are less likely to collapse. There are certain rules they follow. Buildings must have a firm foundation in solid rock rather than soft soil. All parts of the building must be strongly tied together and to the ground. During an earthquake, a building will start to collapse at points where connections are weak. For this reason, steel is used to reinforce the structure. Engineers can make old buildings more earthquake-resistant by adding steel reinforcement beams to these older structures.

Many skyscrapers in San Francisco have been designed to withstand earthquakes. How do engineers accomplish this?

The shape of a building is also important. Box-shaped buildings are stronger than buildings that have a less regular shape. This is because different parts of an irregular-shaped building may sway differently during a quake. This can cause the building to crack and collapse. Skyscrapers and other tall buildings must be designed so they can bend slightly as seismic waves cause them to sway. If a tall building is rigid, the forces set up by swaying would cause it to crack. Engineers sometimes place buildings on special springs that absorb some of the ground movement.

This photograph shows a street in Kobe after the 1995 earthquake. Why is this building lying in the street while the others around it are still standing?

Concept 4.3: Reducing Earthquake Impacts

4.3 | Learn — What should I do to keep myself safe during and after an earthquake?

Activity 14

Design Solutions Like a Scientist

Quick Code: ca4808s

Hands-On Engineering: Earthquake-Resistant House

In this activity, you will **plan**, **build**, **test**, and **modify** a model of a two-story, earthquake-resistant house. You will then **present** your design to the class.

Ask Questions About the Problem

What features should your house include?

_____
_____
_____
_____

What is the best type of foundation to build your house on?

_____
_____
_____
_____
_____
_____

SEP Constructing Explanations and Designing Solutions
SEP Developing and Using Models
CCC Structure and Function

128

What materials do you need? (per group)

- Cardboard (per class)
- Thick rubber bands, 2 (per class)
- Tennis balls (per class)
- Paint stirrer (per class)
- Large binder clips, 2
- Masking tape
- Manila file folder
- Modeling clay
- Toothpicks
- Mini marshmallows, 40–50
- Metric ruler

What Will You Do?

1. **Decide** what the problem is that you are trying to solve.
   _____
   _____

2. **Sketch** your design.

Concept 4.3: Reducing Earthquake Impacts | 129

4.3 | Learn — What should I do to keep myself safe during and after an earthquake?

3. **Build** and **test** your design to determine its effectiveness.
4. **Share** your design with the class, including your diagrams and test results.

Think About the Activity

Do earthquake-resistant houses need different designs from earthquake-resistant office buildings? Explain your answer.

_____
_____
_____
_____
_____
_____

Why are these designs called earthquake-resistant instead of earthquake-proof?

_____
_____
_____
_____
_____
_____
_____

Activity 15
Evaluate Like a Scientist

Quick Code: ca4809s

What Are the Dangers?

Underline the statements that correctly describe earthquake dangers.

Earthquake aftershocks do not cause as much damage as the main shock.

Tsunamis are a serious concern for coastal waters following an earthquake.

Earthquakes can cause highways and bridges to collapse, making evacuation difficult.

Tsunami warnings and procedures in coastal areas are sufficient and well known.

Earthquakes can't be predicted well, even in areas where they are common.

Concept 4.3: Reducing Earthquake Impacts

4.3 | Share — What should I do to keep myself safe during and after an earthquake?

Activity 16
Record Evidence Like a Scientist

Quick Code: ca4810s

The Aftermath of an Earthquake

Now that you have learned about the effects of an earthquake, **watch** the video again. You first saw this in Wonder.

Let's Investigate the Aftermath of an Earthquake

Talk Together

How can you describe the aftermath of earthquakes now? How is your explanation different from before?

SEP Constructing Explanations and Designing Solutions

Look at the Can You Explain? question. You first read this question at the beginning of the lesson.

> **Can You Explain?**
>
> What should I do to keep myself safe during and after an earthquake?

Now, you will use your new ideas about earthquake impacts to answer a question.

1. **Choose** a question. You can use the Can You Explain? question or one of your own. You can also use one of the questions that you wrote at the beginning of the lesson.

   My Question

2. Then, **use** the graphic organizers on the next pages to help you answer the question.

Concept 4.3: Reducing Earthquake Impacts

4.3 | Share — What should I do to keep myself safe during and after an earthquake?

To plan your scientific explanation, first **write** your claim. Your claim is a one-sentence answer to the question you investigated. It answers: What can you conclude? It should not start with yes or no.

My claim:

_____

_____

_____

Data should support your claim. Leave out information that does not support your claim. **List** data that supports your claim.

Data 1

Data 2

Finally, **explain** your reasoning. Reasoning ties together the claim and the evidence. Reasoning shows how or why the data count as evidence to support the claim.

Topic: _____

Data	Reasoning That Supports Claim

4.3 | Share
What should I do to keep myself safe during and after an earthquake?

To keep myself safe during and after an earthquake, I can...

 in Action

 **Activity 17**
Analyze Like a Scientist

Quick Code: ca4811s

Problems and Solutions

Read the text, **watch** the video, and **answer** the questions that follow.

Problems and Solutions

The shaking and tsunamis caused by earthquakes can be dangerous for people and property. How do scientists and engineers work together to keep people safe? Solving the problems caused by earthquakes means deciding where they are most likely to occur. It also means using technology to make people in these places safer.

Seismologists are scientists who study earthquakes. They focus on sensing, recording, measuring, and analyzing data about earthquakes. They know that earthquakes are more likely to occur where Earth's plates meet. They use special sensors to measure and locate earthquakes.

When an earthquake happens, energy travels in seismic waves away from the focus of the earthquake. This is the place in Earth's crust where the earthquake begins. A tool called a seismograph can sense and record these waves, even if they are too weak for people to feel.

SEP Obtaining, Evaluating, and Communicating Information

Knowing the places where strong earthquakes are more likely to occur allows engineers to focus on making buildings in those places safer. Engineers may place rubber shock absorbers in the foundations of large buildings. The rubber absorbs some of the energy from the earthquake. Engineers may use steel bars to brace the walls of the buildings and bridges. Buildings can also be designed to sway and twist without collapsing. New materials and new designs are being developed to make people safer during and after earthquakes. These materials and designs are tested and compared to see which ones are the most effective.

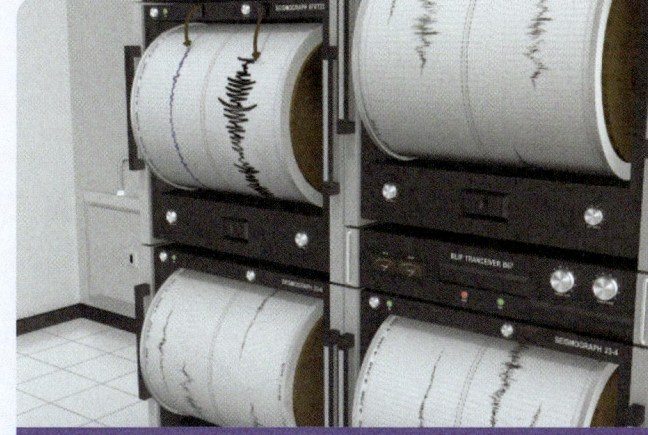

This image shows several seismographs. What does a seismograph do?

Other researchers develop technologies that provide warnings against possible tsunamis. These systems sense earthquakes far out at sea. They send warning messages by satellite to people in nearby areas.

Monitoring Earthquakes

Concept 4.3: Reducing Earthquake Impacts

Tsunami Warning System

Look at the maps. **Circle** the country that could **most** benefit from a tsunami warning system.

Why?

_____
_____
_____
_____
_____

Activity 18

Evaluate Like a Scientist

Quick Code: ca4812s

Review: Reducing Earthquake Impacts

Think about what you have read and seen in this lesson. **Write** down some core ideas you have learned. **Review** your notes with a partner. Your teacher may also have you take a practice test.

_____
_____
_____
_____
_____
_____
_____

Talk Together

Think about what you saw in Get Started. Use your new ideas about reducing earthquake impacts to discuss how earthquakes affect bridge design.

SEP Obtaining, Evaluating, and Communicating Information

Concept 4.3: Reducing Earthquake Impacts

Unit Project

 **Solve Problems Like a Scientist**

Quick Code: ca4816s

Unit Project: Designing a Safe Bridge

In this project, you will use what you know about engineering and technology to design an earthquake-proof bridge.

California's Bay Area has two large cities. San Francisco is on a peninsula on the west side of the bay. Oakland is on the east side of the bay on the mainland. In the early 1930s, people had to travel between the two cities by ferry. When a bridge between the two cities was proposed, some people thought it would never work! The bay has high winds and large waves, and the bedrock the bridge would need to be grounded in is hundreds of feet below the surface.

In 1936, the critics were proved wrong. The San Francisco–Oakland Bay Bridge was built, spanning 8 miles across the bay. The bridge lasted over 50 years without a major problem! Unfortunately, an earthquake in 1989 collapsed a section of the bridge.

The original designers of the bridge had tried to make their bridge earthquake-resistant, but the quake had been too strong. Between 2002 and 2013, a new, more earthquake-resistant bridge was built. How did the designers of the bridge overcome the challenges of building a bridge in the Bay Area?

SEP Constructing Explanations and Designing Solutions
CCC Structure and Function

With your group, **watch** the video. As you watch, **draw** a diagram of your bridge design.

Earthquake-Resistant Bridges

Unit Project

Bridge Design

Brainstorm a design that might improve the bridge even more. Your design can focus on challenges of seismic activity, weather, weight, waves, or something else. **Describe** your new design and explain how you would test it before building it.

_____
_____
_____
_____
_____
_____
_____

Testing a Design

Bridge engineers need to test their designs before they decide on a solution. They are considering each of the following tests:

- Bend and twist the materials the bridge will be built from to see how they respond.
- Build the bridge and wait for an earthquake to see how well it works.
- Create a computer program that will test how the bridge will react to different conditions.
- Make a scale model and see how it responds to earthquake-like movements.

Which of the tests is not a good option for testing how a bridge would perform in an earthquake?

_____
_____

Think About the Activity

Share your design with the class. **Discuss** how well each design solves the problem. **Write** or **draw** your ideas in the chart.

What Worked?	What Didn't Work?

What Could Work Better?

Unit 4: Earthquakes

Grade 4 Resources

- **Bubble Map**
- **Safety in the Science Classroom**
- **Vocabulary Flash Cards**
- Glossary
- Index

Name _____

Bubble Map

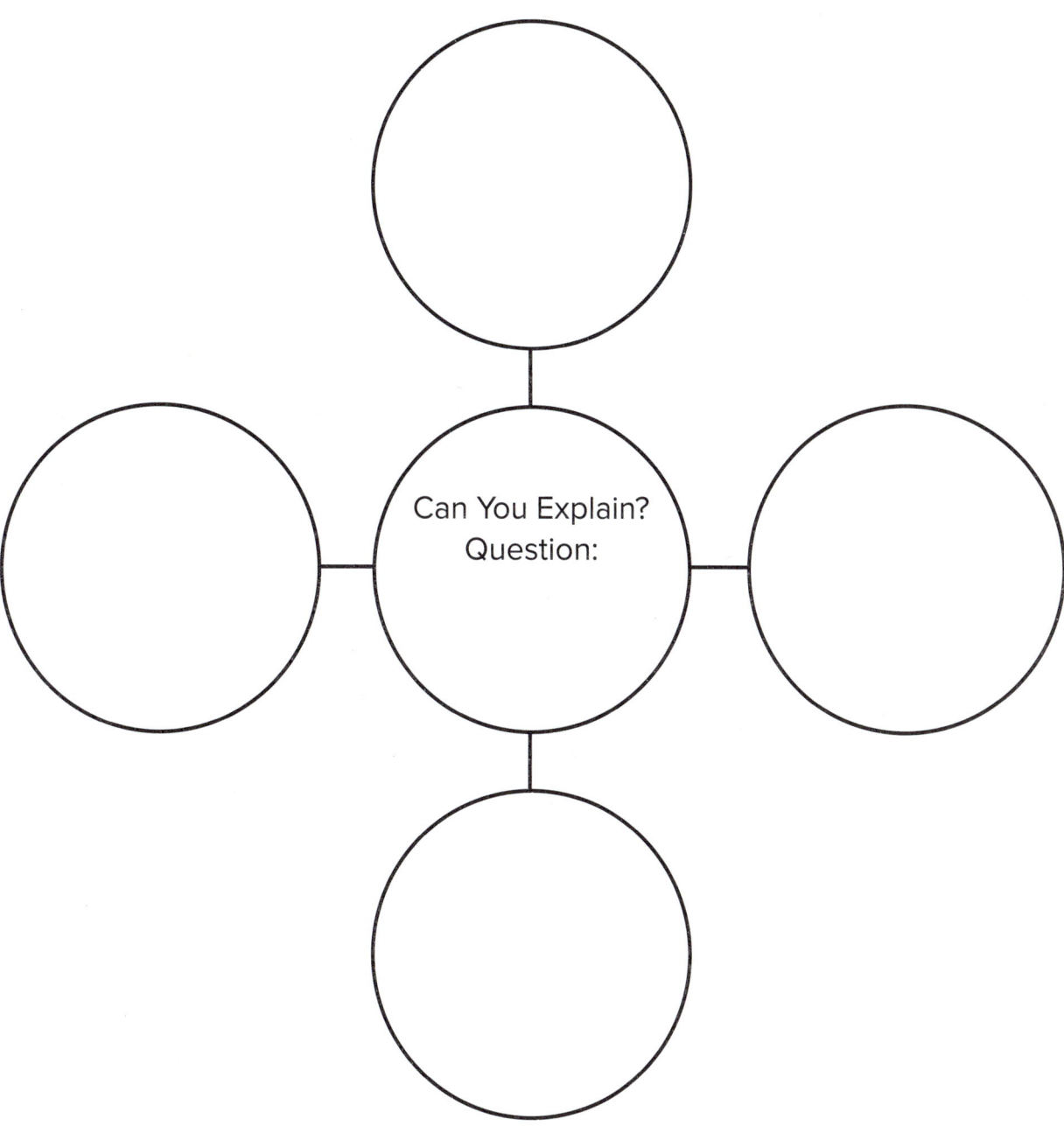

Bubble Map | R3

Safety

Safety in the Science Classroom

Following common safety practices is the first rule of any laboratory or field scientific investigation.

Dress for Safety

One of the most important steps in a safe investigation is dressing appropriately.

- Splash goggles need to be kept on during the entire investigation.
- Use gloves to protect your hands when handling chemicals or organisms.
- Tie back long hair to prevent it from coming in contact with chemicals or a heat source.
- Wear proper clothing and clothing protection. Roll up long sleeves, and if they are available, wear a lab coat or apron over your clothes. Always wear close toed shoes. During field investigations, wear long pants and long sleeves.

Be Prepared for Accidents

Even if you are practicing safe behavior during an investigation, accidents can happen. Learn the emergency equipment location in your classroom and how to use it.

- The eye and face wash station can help if a harmful substance or foreign object gets into your eyes or onto your face.
- Fire blankets and fire extinguishers can be used to smother and put out fires in the laboratory. Talk to your teacher about fire safety in the lab. He or she may not want you to directly handle the fire blanket and fire extinguisher. However, you should still know where these items are in case the teacher asks you to retrieve them.
- Most importantly, when an accident occurs, immediately alert your teacher and classmates. Do not try to keep the accident a secret or respond to it by yourself. Your teacher and classmates can help you.

Practice Safe Behavior

There are many ways to stay safe during a scientific investigation. You should always use safe and appropriate behavior before, during, and after your investigation.

Safety Goggles

- Read the all of the steps of the procedure before beginning your investigation. Make sure you understand all the steps. Ask your teacher for help if you do not understand any part of the procedure.

- Gather all your materials and keep your workstation neat and organized. Label any chemicals you are using.

- During the investigation, be sure to follow the steps of the procedure exactly. Use only directions and materials that have been approved by your teacher.

- Eating and drinking are not allowed during an investigation. If asked to observe the odor of a substance, do so using the correct procedure known as wafting, in which you cup your hand over the container holding the substance and gently wave enough air toward your face to make sense of the smell.

- When performing investigations, stay focused on the steps of the procedure and your behavior during the investigation. During investigations, there are many materials and equipment that can cause injuries.

- Treat animals and plants with respect during an investigation.

- After the investigation is over, appropriately dispose of any chemicals or other materials that you have used. Ask your teacher if you are unsure of how to dispose of anything.

- Make sure that you have returned any extra materials and pieces of equipment to the correct storage space.

- Leave your workstation clean and neat. Wash your hands thoroughly.

Vocabulary Flash Cards

air

the part of the atmosphere closest to Earth; the part of the atmosphere that organisms on Earth use for respiration

amplitude

height or "strength" of a wave

behavior

all of the actions and reactions of an animal

earthquake

a sudden shaking of the ground caused by the movement of rock underground

energy

Image: Paul Fuqua

the ability to do work or cause change; the ability to move an object some distance

engineer

Image: Nino Care/Pixabay

engineers have special skills; they design things that help solve problems

fault

Image: Beth Read/Shutterstock

a crack in a body of rock that the rock can move along

forecast

Image: RitaE/Pixabay

(n) a prediction about what the weather will be like in the future based on weather data

light

waves of electromagnetic energy; electromagnetic energy that people can see

magma

melted rock located beneath Earth's surface

magnetic field

a region in space near a magnet or electric current in which magnetic forces can be detected

model

a drawing, object, or idea that represents a real event, object, or process

mountain

an area of land that forms a peak at a high elevation

predict

to guess what will happen in the future

seismic

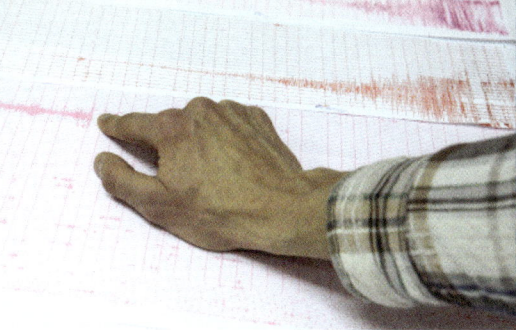

having to do with earthquakes or earth vibrations

seismic wave

waves of energy that travel through Earth's interior due to an earthquake, other tectonic forces, or an explosion

sound

Image: Paul Fuqua

waves of energy that travel through Earth's interior due to an earthquake, other tectonic forces, or an explosion

sound wave

Image: radioshoot/Shutterstock

a sound vibration as it is passing through a material: Most sound waves spread out in every direction from their source.

system

Image: Gil Meshulam/Shutterstock

a group of related objects that work together to perform a function

tectonic plate

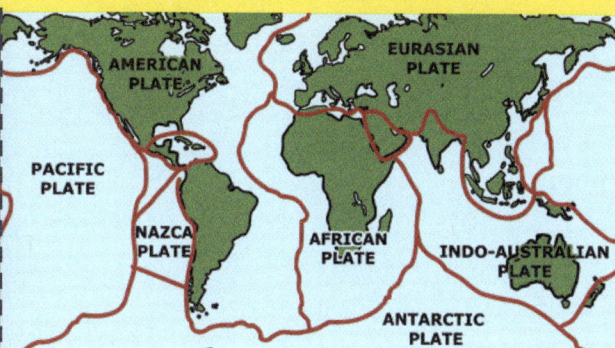

Image: Discovery Communications, Inc.

one of several huge pieces of Earth's crust

Vocabulary Flash Cards | R15

tsunami

a giant ocean wave

valley

a low area of land between two higher areas, often formed by water

wave

a disturbance caused by a vibration; Waves travel away from the source that makes them.

wavelength

the distance between one peak and the next on a wave

Vocabulary Flash Cards | R17

Glossary

English ——— A ——— Español

acceleration
to increase speed

aceleración
aumentar la velocidad

adaptation
something a plant or animal does to help it survive in its environment (related word: adapt)

adaptación
proceso mediante el cual las características de una especie cambian a través de varias generaciones como respuesta al medio ambiente (palabra relacionada: adaptar)

air
the part of the atmosphere closest to Earth; the part of the atmosphere that organisms on Earth use for respiration

aire
parte de la atmósfera más cercana a la Tierra; la parte de la atmósfera que los organismos que habitan la Tierra utilizan para respirar

amplitude
height or "strength" of a wave

amplitud
altura o "fuerza" de una onda

analog
one continuous signal that does not have any breaks

analógico
una señal continua que no tiene ninguna interrupción

antenna
a device that receives radio waves and television signals

antena
un dispositivo que recibe ondas de radio y señales de televisión

Arctic
being from an icy climate, such as the north pole

ártico
que tiene relación con el Polo Norte o el área que lo rodea

--- **B** ---

behavior
all of the actions and reactions of an animal or a person (related word: behave)

conducta
todas las acciones y reacciones de un animal (palabra relacionada: comportarse)

brain
the main control center in an animal body; part of the central nervous system

cerebro
principal centro de control en el cuerpo de un animal; parte del sistema nervioso central

--- **C** ---

camouflage
the coloring or patterns on an animal's body that allow it to blend in with its environment

camuflaje
color o apariencia del cuerpo de un animal que le permite mezclarse con su medioambiente

canyon
a deep valley carved by flowing water

cañón
valle profundo esculpido por el flujo del agua

chemical energy
energy that can be changed into motion and heat

energía química
energía que está almacenada en las cadenas entre átomos

chemical weathering
changes to rocks and minerals on Earth's surface that are caused by chemical reactions

meteorización química
cambios en las rocas y minerales de la superficie de la Tierra causados por reacciones químicas

code
a way to communicate by sending messages using dots and dashes

código
una forma de comunicarse enviando mensajes con puntos y rayas

collision
the moment where two objects hit or make contact in a forceful way

colisión
el momento en el que dos objetos chocan o hacen contacto de forma contundente

conduction
when energy moves directly from one object to another

conducción
cuando la energía se mueve directamente de un objeto a otro

conservation of energy
energy can not be created or destroyed, it can only be changed from one form to another, such as when electrical energy is changed into heat energy

conservación de la energía
la energía no se puede crear o destruir; solo se puede cambiar de una forma a otra, como cuando la energía eléctrica cambia a energía térmica

conserve
to protect something, or prevent the wasteful overuse of a resource

conservar
proteger algo o evitar el uso excesivo e ineficiente de un recurso

convert (v)
to change forms

convertir (v)
cambiar de forma

D

delta
a fan-shaped mass of mud and other sediment that forms where a river enters a large body of water

delta
masa de barro y otros sedimentos parecida a un abanico, que se forma donde un río ingresa a un gran cuerpo de agua

deposition
laying sediment back down after erosion moves it around

sedimentación
volver a depositar sedimentos una vez que la erosión los arrastra

digestive system
the body system that breaks down food into tiny pieces so that the body's cells can use it for energy

sistema digestivo
sistema del cuerpo que divide los alimentos en pequeños trozos para que las células del cuerpo puedan usarlos para obtener energía.

digital
a signal that is not continuous and is made up of tiny separate pieces

digital
una señal que no es continua y está compuesta por diminutas partes separadas

disease
a condition that disrupts processes in the body and usually causes an illness

enfermedad
afección que perturba los procesos del cuerpo

dune
a hill of sand created by the wind

duna
colina de arena creada por el viento

E

ear
organ for hearing

oído
órgano para oir

Earth
the third planet from the sun; the planet on which we live (related words: earthly; earth – meaning soil or dirt)

Tierra
tercer planeta desde el Sol; planeta en el cual vivimos (palabras relacionadas: terrenal; tierra en el sentido de suelo o suciedad)

earthquake
a sudden shaking of the ground caused by the movement of rock underground

terremoto
repentina sacudida de la tierra causada por el movimiento de rocas subterráneas

ecosystem
all the living and nonliving things in an area that interact with each other

ecosistema
todos los seres vivos y objetos sin vida de un área, que se interrelacionan entre sí

electromagnetic spectrum
the full range of frequencies of electromagnetic waves

espectro electromagnético
rango completo de frecuencias de las ondas electromagnéticas

energy
the ability to do work or cause change; the ability to move an object some distance

energía
habilidad para hacer un trabajo o producir un cambio; habilidad para mover un objeto cierta distancia

energy source
where a form of energy begins

fuente de energía
donde comienza una forma de energía

Glossary | R25

energy transfer
the transfer of energy from one organism to another through a food chain or web; or the transfer of energy from one object to another, such as heat energy

transferencia de energía
transmisión de energía desde un organismo a otro a través de una cadena o red de alimentos; o transferencia de energía desde un objeto a otro, como por ejemplo la energía del calor

engineer
Engineers have special skills. They design things that help solve problems.

ingeniero
Los ingenieros poseen habilidades especiales. Diseñan cosas que ayudan a resolver problemas.

environment
all the living and nonliving things that surround an organism

medio ambiente
todos los seres vivos y objetos sin vida que rodean a un organismo

erosion
the removal of weathered rock material. After rocks have been broken down, the small particles are transported to other locations by wind, water, ice, and gravity

erosión
proceso por el cual el viento, el agua, el hielo y otras cosas mueven trozos de roca y suelo sobre la superficie de la Tierra

erupt
the action of lava coming out of a hole or crack in Earth's surface; the sudden release of hot gasses or lava built up inside a volcano (related word: eruption)

erupción
acción de la lava que sale de un agujero o cráter de la superficie de la Tierra; repentina liberación de gases o lava calientes formados en el interior de un volcán (palabra relacionada: erupción)

extinct
describes a species of animals that once lived on Earth but which no longer exists (related word: extinction)

extinto
palabra que hace referencia a una especie de animales que una vez habitó la Tierra, pero que ya no existe (palabra relacionada: extinción)

--- F ---

fault
a fracture, or a break, in the Earth's crust (related word: faulting)

falla
grieta en el cuerpo de una roca que hace que la roca se desplace (palabra relacionada: fallas)

feature
things that describe what something looks like

rasgo
cosas que describen cómo se ve algo

force
a pull or push that is applied to an object

fuerza
acción de atraer o empujar que se aplica a un objeto

forecast
(v) to analyze weather data and make an educated guess about weather in the future; (n) a prediction about what the weather will be like in the future based on weather data

pronosticar / pronóstico
(v) analizar los datos del tiempo y hacer una conjetura informada sobre el tiempo en el futuro; (s) predicción sobre cómo será el tiempo en el futuro en base a datos

fossil fuels
fuels that come from very old life forms that decomposed over a long period of time, like coal, oil, and natural gas

combustibles fósiles
varios tipos de combustibles formados de manera natural a partir de los restos de plantas y animales que murieron hace miles o millones de años

friction
a force that slows down or stops motion

fricción
fuerza que se opone al movimiento de un cuerpo sobre una superficie o a través de un gas o un líquido

fuel
any material that can be used for energy

combustible
todo material que puede usarse para producir energía

--- G ---

generate
to produce by turning a form of energy into electricity

generar
producir convirtiendo una forma de energía en electricidad

geothermal
heat found deep within Earth

geotérmica
calor que se encuentra en la profundidad de la Tierra

glacier
a large sheet of ice or snow that moves slowly over Earth's surface

glaciar
ran sábana de hielo o nieve que se mueve lentamente sobre la superficie de la Tierra

gravitational potential energy
energy stored in an object based on its height and mass

energía potencial gravitacional
energía almacenada debida a la ubicación en un campo gravitacional

gravity
the force that pulls an object toward the center of Earth (related word: gravitational)

gravedad
fuerza que existe entre dos objetos cualquiera que tienen masa (palabra relacionada: gravitacional)

H

heart
the muscular organ of an animal that pumps blood throughout the body

corazón
órgano muscular de un animal que bombea sangre a través del cuerpo

heat
the transfer of thermal energy

calor
transferencia de energía térmica

hibernate
to reduce body movement during the winter in an effort to conserve energy (related word: hibernation)

hibernar
reducir el movimiento del cuerpo durante el invierno con la finalidad de conservar la energía (palabra relacionada: hibernación)

I

information
facts or data about something; the arrangement or sequence of facts or data

información
hechos o datos sobre algo; la organización o secuencia de hechos o datos

K

kinetic energy
the energy an object has because of its motion

energía cinética
energía que posee un objeto a causa de su movimiento

L

landform
a large natural structure on Earth's surface, such as a mountain, a plain, or a valley

accidente geográfico
estructura natural grande que se encuentra en la superficie de la Tierra, como una montaña, una llanura o un valle

lava
molten rock that comes through holes or cracks in Earth's crust that may be a mixture of liquid and gas but will turn into solid rock once cooled

lava
roca fundida que sale por orificios o grietas en la corteza terrestre, y que puede ser una mezcla de líquido y gas pero se convierte en roca sólida al enfriarse

light
a form of energy that moves in waves and particles and can be seen

luz
ondas de energía electromagnéticas; energía electromagnética que la gente puede ver

M

magma
melted rock located beneath Earth's surface

magma
roca fundida que se encuentra debajo de la superficie de la Tierra

magnetic field
a region in space near a magnet or electric current in which magnetic forces can be detected

campo magnético
región en el espacio cerca de un imán o de una corriente eléctrica, donde pueden detectarse fuerzas magnéticas

map
a flat model of an area

mapa
modelo plano de un área

mass
the amount of matter in an object

masa
cantidad de materia en un objeto

matter
material that has mass and takes up some amount of space

materia
material que tiene masa y ocupa cierta cantidad de espacio

meander
winding or indirect movement or course

meandro
movimiento o curso serpenteante o indirecto

Glossary | R33

migration
the movement of a group of organisms from one place to another, usually due to a change in seasons

migración
movimiento de un grupo de organismos de un lugar a otro, generalmente debido a un cambio de estaciones

model
a drawing, object, or idea that represents a real event, object, or process

modelo
dibujo, objeto o idea que representa un evento, objeto, o proceso real

motion
when something moves from one place to another (related words: move, movement)

movimiento
cambio en la posición de un objeto en comparación con otro objeto (palabra relacionada: mover, desplazamiento)

mountain
an area of land that forms a peak at a high elevation (related term: mountain range)

montaña
área de tierra que forma un pico a una elevación alta (palabra relacionada: cadena montañosa)

N

nerve
a cell of the nervous system that carries signals to the body from the brain, and from the body to the brain and/or spinal cord

nervio
célula del sistema nervioso que lleva señales al cuerpo desde el cerebro, y desde el cuerpo al cerebro y/o médula espinal

nonrenewable
once it is used, it cannot be made or reused again

no renovable
no renovable

nonrenewable resource
a natural resource of which a finite amount exists, or one that cannot be replaced with currently available technologies

recurso no renovable
recurso natural del cual existe una cantidad finita, o uno que no puede remplazarse con las tecnologías actualmente disponibles

nuclear energy
the energy released when the nucleus of an atom is split apart or combined with another nucleus

energía nuclear
energía liberada cuando el núcleo de un átomo se divide o combina con otro átomo

O

ocean
a large body of salt water that covers most of Earth

océano
gran cuerpo de agua salada que cubre la mayor parte de la Tierra

opaque
describes an object that light cannot travel through

opaco
describe un objeto que la luz no puede atravesar

organ
a group of tissues that performs a complex function in a body

órgano
conjunto de tejidos que realizan una función compleja en el cuerpo

organism
any individual living thing

organismo
todo ser vivo individual

P

photosynthesis
the process in which plants and some other organisms use the energy in sunlight to make food

fotosíntesis
proceso en el cual las plantas y algunos otros organismos usan la energía del Sol para producir alimentos

pollute
to put harmful materials into the air, water, or soil (related words: pollution, pollutant)

contaminar
poner materiales perjudiciales en el aire, agua o suelo (palabras relacionadas: contaminación, contaminante)

---

pollution
when harmful materials have been put into the air, water, or soil (related word: pollute)

contaminación
cuando se introducen materiales perjudiciales en el aire, el agua o el suelo (palabra relacionada: contaminar)

---

potential energy
the amount of energy that is stored in an object; energy that an object has because of its position relative to other objects

energía potencial
cantidad de energía almacenada en un objeto; energía que tiene un objeto debido a su posición relativa con otros objetos

---

predator
an animal that hunts and eats another animal

depredador
animal que caza y come a otro animal

predict
to guess what will happen in the future (related word: prediction)

predecir
adivinar qué sucederá en el futuro (palabra relacionada: predicción)

prey
an animal that is hunted and eaten by another animal

presa
animal que es cazado y comido por otro

pupil
the black circle at the center of an iris that controls how much light enters the eye

pupila
círculo negro en el centro del iris que controla cuánta luz entra al ojo

R

radiant energy
energy that does not need matter to travel; light

energía radiante
energía que no necesita de la materia para viajar; luz

radiation
electromagnetic energy (related word: radiate)

radiación
energía electromagnética (palabra relacionada: irradiar)

receptor
nerves located in different parts of the body that are especially adapted to receive information from the environment

receptor
nervios ubicados en diferentes partes del cuerpo que están especialmente adaptados para recibir información del medio ambiente

reflect
light bouncing off a surface (related word: reflection)

reflejar
golpear sobre una superficie y rebotar en la dirección opuesta (palabra relacionada: reflexión)

reflex
an automatic response

reflejo
respuesta automática

refract
to bend light as it passes through a material (related word: refraction)

refractar
inclinación de la luz cuando pasa a través de un material (palabra relacionada: refracción)

remote (adj)
to be operated from a distance

remoto (adj)
que se opera a distancia

renewable
to reuse or make new again

renovable
reutilizar o volver a hacer de nuevo

renewable resource
a natural resource that can be replaced

recurso renovable
recurso natural que puede reemplazarse

reproduce
to make more of a species; to have offspring (related word: reproduction)

reproducir
hacer más de una especie; tener descendencia (palabra relacionada: reproducción)

resistance
when materials do not let energy transfer through them

resistencia
cuando los materiales no permiten la transferencia de energía a través de ellos

resource
a naturally occurring material in or on Earth's crust or atmosphere of potential use to humans

recurso
material que se origina de forma natural en o sobre la corteza o la atmósfera de la Tierra, que es de uso potencial para los seres humanos

rotate
turning around on an axis; spinning (related word: rotation)

rotar
girar sobre un eje; dar vueltas (palabra relacionada: rotación)

---- S ----

satellite
a natural or artificial object that revolves around another object in space

satélite
objeto natural o artificial que gira alrededor de otro objeto en el espacio

sediment
solid material, moved by wind and water, that settles on the surface of land or the bottom of a body of water

sedimento
material sólido que el viento o el agua mueve y que se asienta en la superficie de la tierra o en el fondo de un cuerpo de agua

seismic
having to do with earthquakes or earth vibrations

sísmico
relativo a los terremotos o a las vibraciones de la Tierra

seismic wave
waves of energy that travel through the Earth

onda sísmica
ondas de energía que viajan a través del interior de la Tierra debido a un terremoto, otras fuerzas tectónicas o una explosión

senses
taste, touch, sight, smell, and hearing (related word: sensory)

sentidos
gusto, tacto, visión, olfato y audición (palabra relacionada: sensorial)

skin
an organ that covers and protects the bodies of many animals

piel
órgano que cubre y protege los cuerpos de muchos animales

soil
the outer layer of Earth's crust in which plants can grow; made of bits of dead plant and animal material as well as bits of rocks and minerals

suelo
capa externa de la corteza de la Tierra en donde crecen las plantas; formada por pedazos de plantas y animales muertos, así como por pedazos de rocas y de minerales

solar energy energy that comes from the sun	**energía solar** energia que proviene del Sol
sound anything you can hear that travels by making vibrations in air, water, and solids	**sonido** vibración que viaja a través de un material, como el aire o el agua; lo que se percibe a través de la audición
sound wave a sound vibration as it is passing through a material: Most sound waves spread out in every direction from their source.	**onda sonora** vibración de sonido que se produce cuando se atraviesa un material: la mayoría se dispersa desde la fuente en todas direcciones.
speed the measurement of how fast an object is moving	**velocidad** distancia recorrida por unidad de tiempo
stimulus things in the environment that cause us to react or have a physical response	**estímulo** algo en el medio ambiente que nos hace reaccionar o tener una respuesta física

stomach
a muscular organ in the body where chemical and mechanical digestion take place

estómago
órgano muscular del cuerpo donde tiene lugar la digestión química y mecánica

sun
any star around which planets revolve

sol
toda estrella alrededor de la cual giran los planetas

survive
to continue living or existing: an organism survives until it dies; a species survives until it becomes extinct (related word: survival)

sobrevivir
continuar viviendo o existiendo: un organismo sobrevive hasta que muere; una especie sobrevive hasta que se extingue (palabra relacionada: supervivencia)

system
a group of related objects that work together to perform a function

sistema
grupo de objetos relacionados que funcionan juntos para realizar una función

T

tectonic plate
one of several huge pieces of Earth's crust

placa tectónica
una de las muchas piezas enormes de la corteza terrestre

thermal energy
energy in the form of heat

energía térmica
energía en forma de calor

tongue
an organ in the mouth that helps in eating and speaking

lengua
órgano de la boca que ayuda a comer y hablar

topographic map
a map that shows the size and location of an area's features such as vegetation, roads, and buildings

mapa topográfico
mapa que muestra el relieve y otras características de un área

trait
a characteristic or property of an organism

rasgo
característica o propiedad de un organismo

transparent
describes materials through which light can travel; materials that can be seen through

transparente
describe materiales a través de los cuales puede viajar la luz; materiales a través de los cuales se puede ver

tsunami
a giant ocean wave (related word: tidal wave)

tsunami
hola oceánica gigante (palabra relacionada: maremoto)

--- V ---

valley
a low area of land between two higher areas, often formed by water

valle
área baja de tierra entre dos áreas más altas, generalmente formada por el agua

volcano
an opening in Earth's surface through which magma and gases or only gases erupt (related word: volcanic)

volcán
abertura en la superficie de la Tierra a través de la cual el magma y los gases o sólo los gases hacen erupción (palabra relacionada: volcánico)

water
a compound made of hydrogen and oxygen; can be in either a liquid, ice, or vapor form and has no taste or smell

agua
compuesto formado por hidrógeno y oxígeno

wave
a disturbance caused by a vibration; waves travel away from the source that makes them

onda
perturbación causada por una vibración que se aleja de la fuente que la forma

wavelength
the distance between one peak and the next on a wave

longitud de onda
distancia entre un pico y otro en una onda

weathering
the physical or chemical breakdown of rocks and minerals into smaller pieces or aqueous solutions on Earth's surface

meteorización
desintegración física o química de rocas y minerales en pedazos más pequeño o en soluciones acuosas en la superficie de la Tierra

work
a force applied to an object over a distance

trabajo
fuerza aplicada a un objeto a lo largo de una distancia

Index

A

Air 65
Amplitude
 and wavelength 67–68
Analyze Like a Scientist 16–17, 26, 27–29, 32–34, 35–37, 43–46, 55–56, 65–66, 67–68, 88–91, 100–101, 107–109, 110–113, 118–123, 125–127, 138–140
Ask Questions Like a Scientist 10–11, 52–54, 98–99

B

Behavior 28

C

Can You Explain? 8, 39, 50, 82, 96, 133
Continents
 and the sea floor 25

E

Earthquakes 2
 and aftershocks 108
 and faults 16
 and tsunami 109
 change earth's surface 32
 distribution of 26
 experiencing 10
 loss of life from 100–101
 predicting 27–28
 preparation 107–109, 110–112
 resistant buildings 125–127
 surviving 118
 what happens during? 12
 zones 30

Energy 55, 65, 67–68
Engineer 119–122, 125–127, 138–139, 142, 144
Evaluate Like a Scientist 14–15, 30–31, 47, 58–60, 80, 92, 102–109, 131, 141
Design Solutions Like a Scientist 72–75, 115–117, 128–130

F

Faults 16–17, 27, 33, 43– 45, 55
Forecast 28

G

Geology 43–45

H

Hands-On Activities 72–75, 115–117, 128–130

L

Light 65

M

Magnetic field 28
Model 121
Mountain 32, 33

O

Observe Like a Scientist 12–13, 25, 57, 64, 70–71, 76, 77, 78–79, 106, 114, 124

P

Predict 28

R

Record Evidence Like a Scientist 38–42, 82–86, 132–135

S

Seismic 108, 144
Seismic waves 49, 65, 68
    recording 70
Seismology 43, 88–89
Shockwave 57
Solve Problems Like a Scientist 4–5, 142–145
Sound 65
Sound wave 65
STEM in Action 43–46, 88–91, 138–140
System 95, 107

T

Tectonic plates 26, 30, 43, 142
Think Like a Scientist 18–24, 62–64, 104–105

Tsunami
    and earthquakes 109
    researching 138–139
    warning system 80

U

Unit Project 4–5, 142–145

V

Valley 32

W

Wavelength 67–68
Waves
    amplitude 67–68
    and energy 55
    mediums and 64
    movement 65–66